Copyright © 2017 by Ryan L. Jones

All rights reserved.
Published in the United States by Ryan L. Jones

Copyright for text by Ryan L. Jones
Copyright for illustrations by Ryan L. Jones, Marilyn Calza, Chris Lynch, Brandi Wynn, and Jona Monet

ISBN 978-0-692-97549-7
ebook ISBN 978-0-692-97550-3

First Edition

Written by
 Ryan L. Jones

Designed by
 Mayilyn Calza and Ryan L. Jones

Illustrated by
 Chris Lynch, Marilyn Calza, Brandi Wynn, and Jona Monet

IT'S NOT TAKING THE HIGH ROAD, IT'S THE WAY.

The atom is what the atom is.
Some say that's where it begins.

Still the atom is what the atom is,
Just buzzing around with its friends.

As of now, this is all I'll say,
But you'll come back and tell me more one day.

"One day!?" You might ask…
"tell you more!?" You may think…
But don't you fret or worry
Because that day will be here in just a blink.

As we rise
We shall see
Something really cool
Called the molecule.
From the molecule you'll see the mystical
and magical DNA,
which will open up worlds more of play.

Play on, young one, play on…
It's in these moments that we best create.
Learning more and more, you will find…

THE CODED STAIRCASE CALLED DNA IS QUITE DIVINE, AND IN EVERY CELL OF WHICH WE'RE MADE...

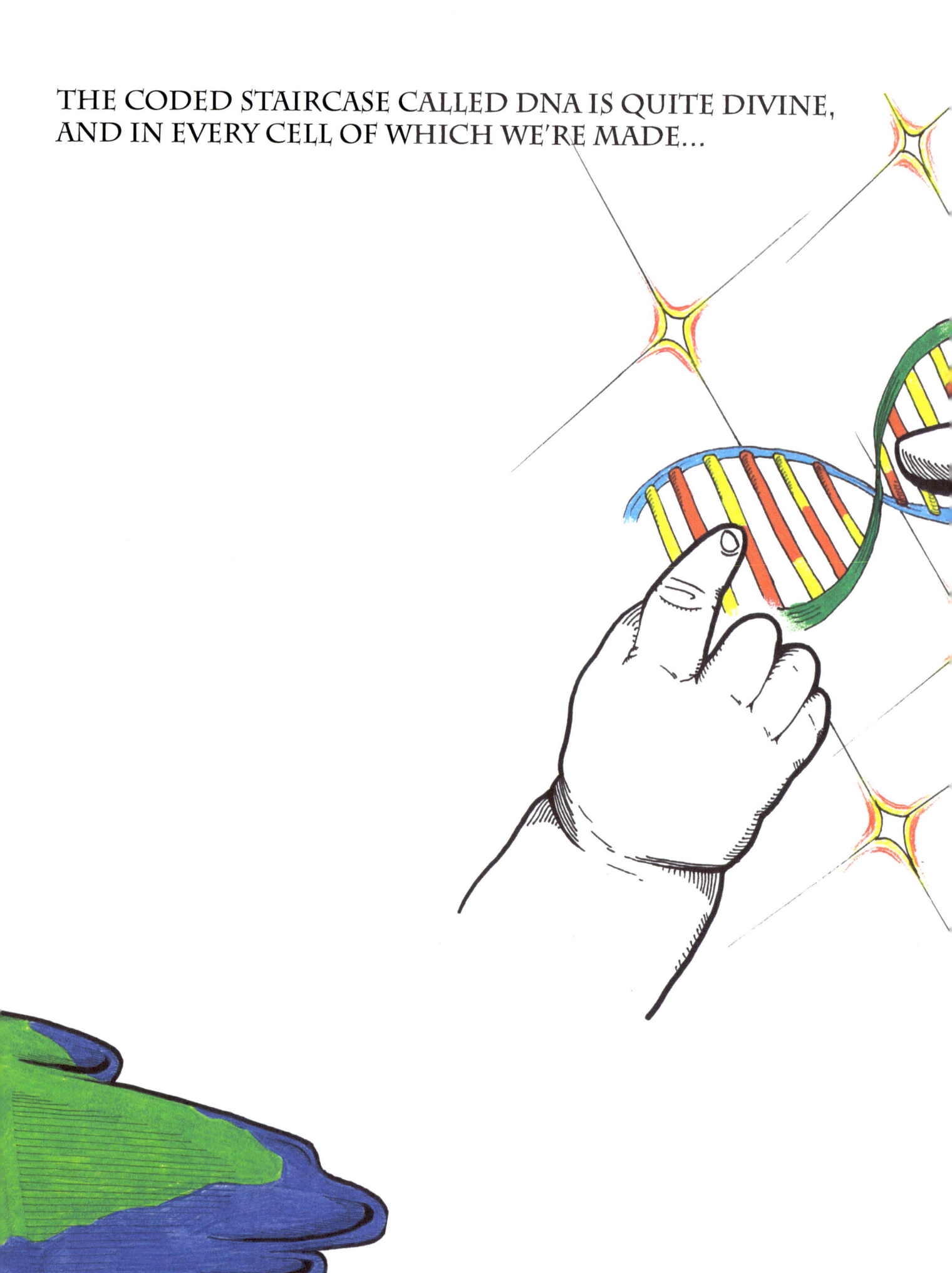

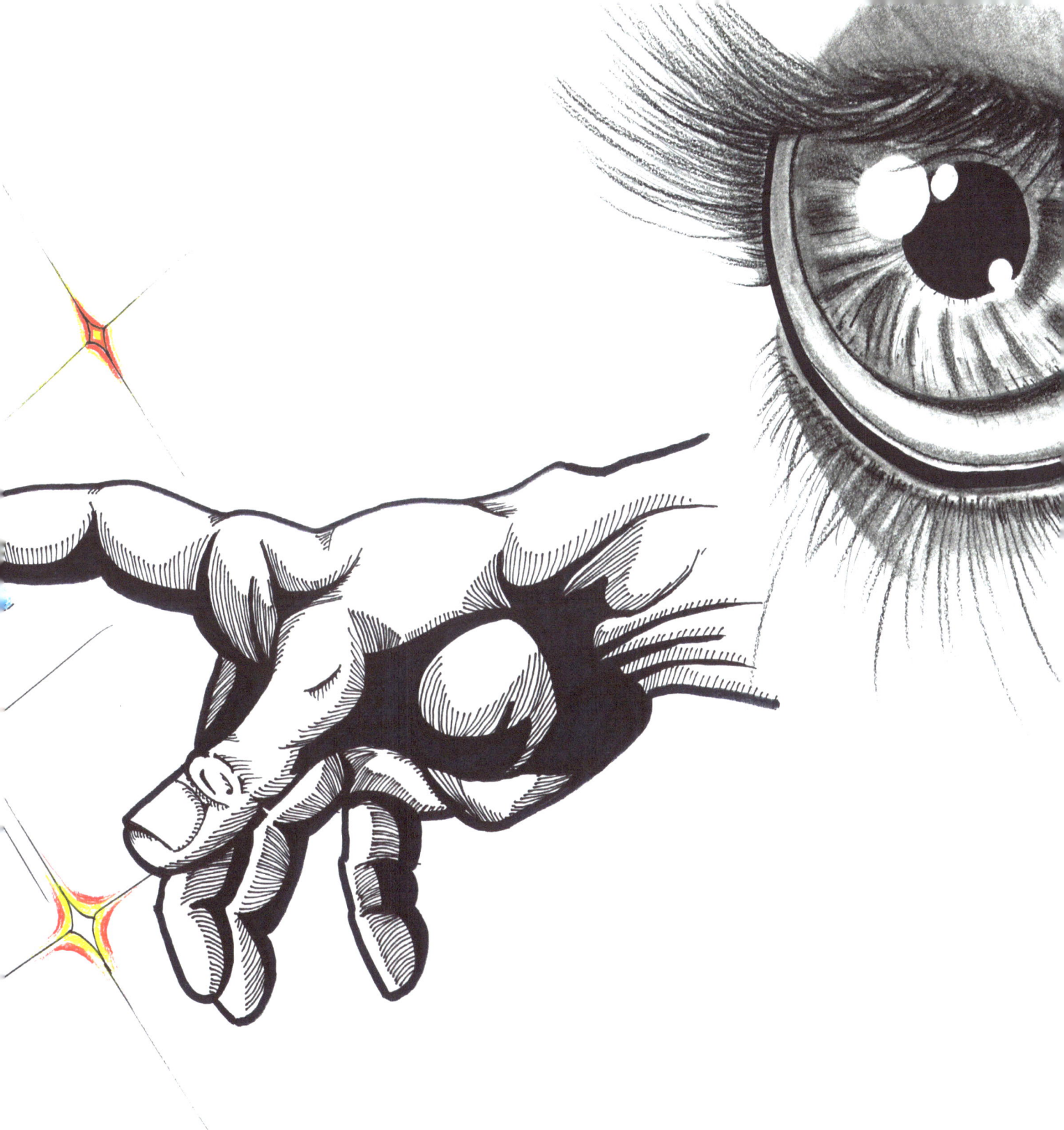

Of which we're made is of many things,
And all you'll know will come from how you choose to grow.

Ever up the staircase of learning and knowledge you shall climb,

you will see there is way more than this simple rhyme.

e birds, the trees, the moon, the stars out in space, you should find this to be
nost wondrous place.

All things are both near and far,
Just depends on how you look,
Kind of like you look in this book.

Rainbows and lilies are made by the same,
With a different result, the Sun radiates its fame.
So pleased is the Sun,
Giving and creating life and things.
You too can share your light
To help yourself and others grow as you go.

Here on Earth we go round n' round,
Remember that as you playfully roll on the ground,
That's because the Earth's dance (see the
Electric Slide) with others make it allowed.

Explore the world,
There are many bumps and grooves,
Don't worry too much because it can get really smooth.

Climb this magical staircase of life,
You have already and will.
Be sure to respect all life,
From the elephant to the krill.

"Why is that!?" you might inquire,
It's because we all have the same desire.
We all want to aspire...for higher.
All made up of atoms and cells.
That should ring a bell.

There are cells in the heart
And cells in the brain,
They have different responsibilities and functions
But ultimately they are the same.
Cells just being cells
Doing their thing.

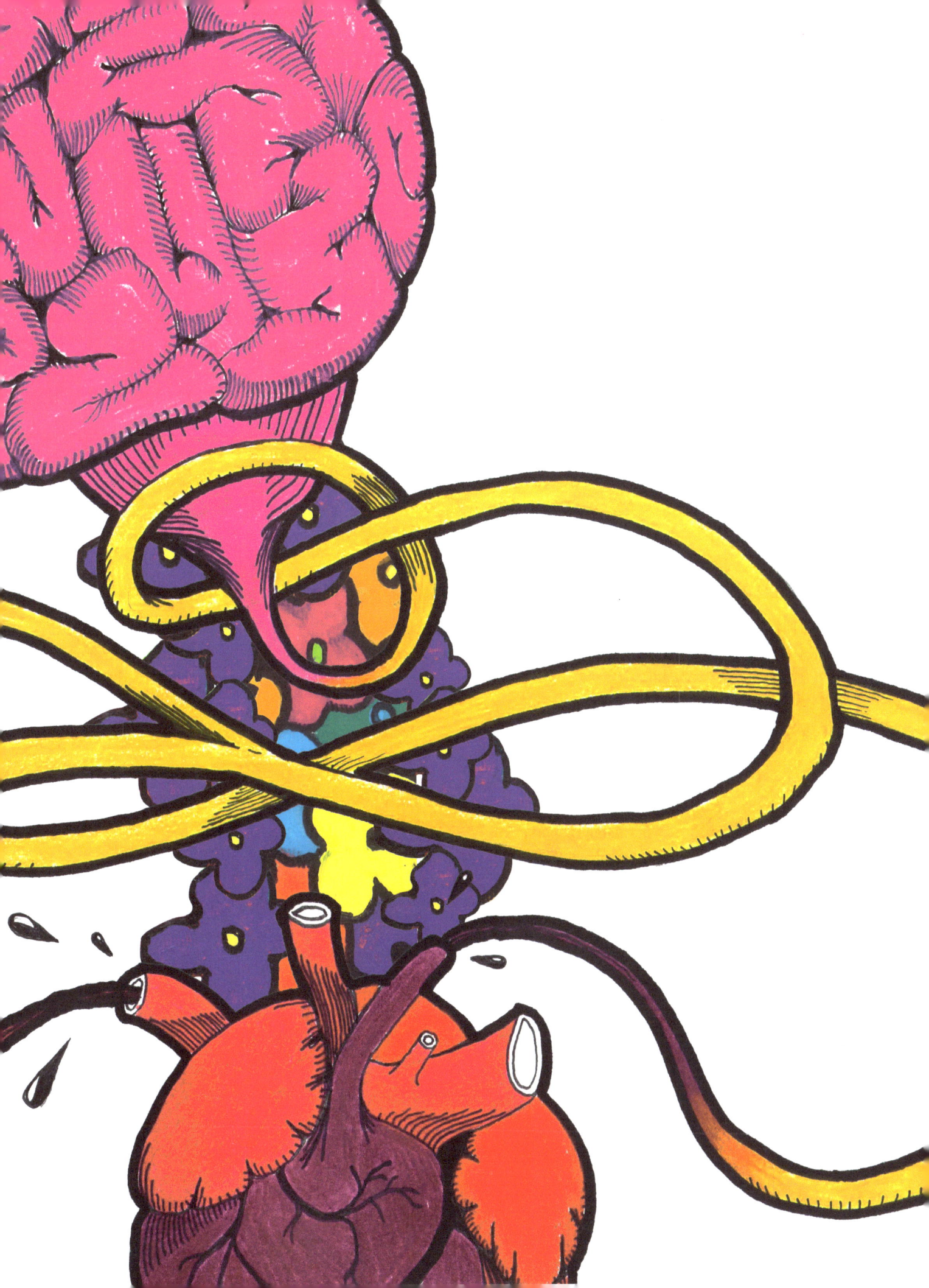

CELLS IN YOUR NOSE
AND CELLS IN YOUR SKIN
CERTAINLY CELLS DON'T CARE ABOUT
THE SKIN THAT YOU'RE IN.

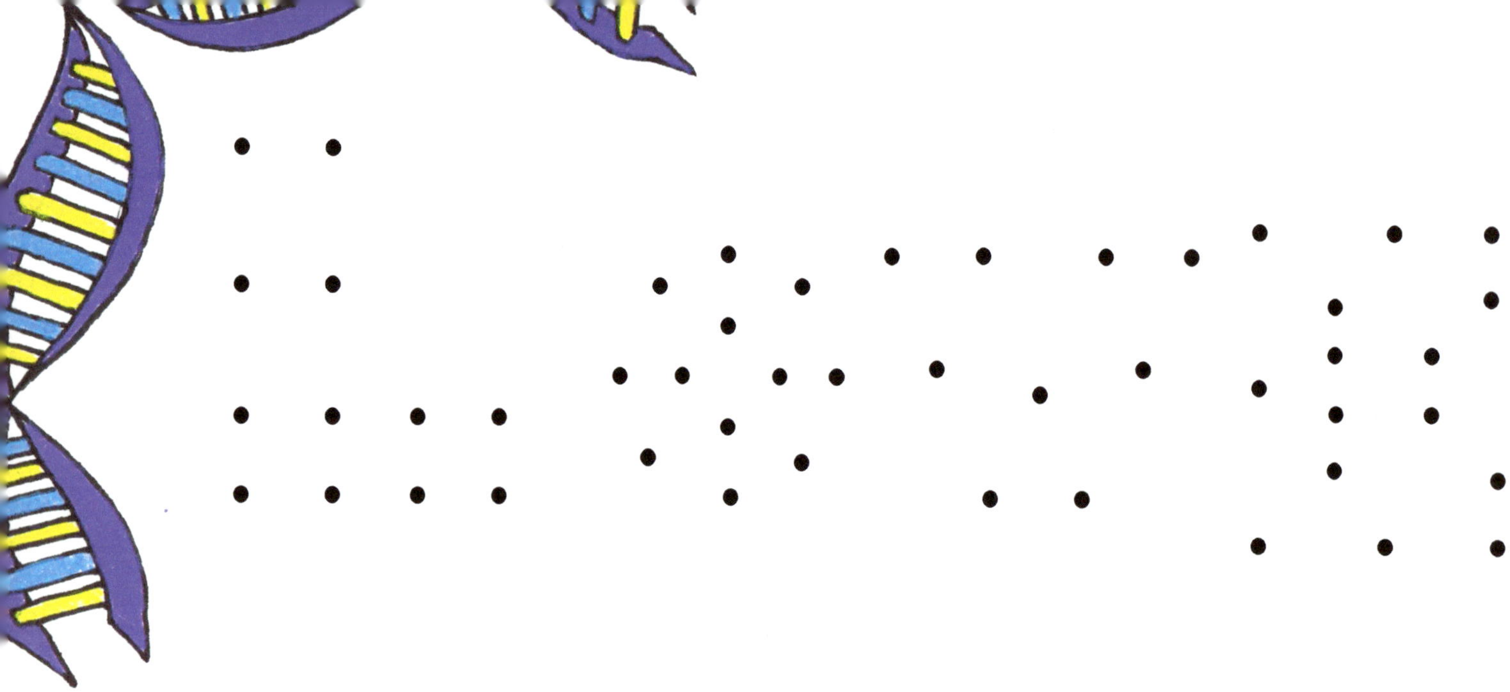

Living beings come in different shapes and sizes,
Some with more cells and some with less,
But no cell really cares, they just give it their best!
Creating your size and shape as they are told by your DNA,
And from the DNA looking down and down,
We can find where life begins...
It's the atom buzzing around with its friends.

So look up with your eyes,
And see the universe's beauty,
Dream big...dream bigger...dream more...dream forever...
And create with love, like it is your duty.

As you can see, there are ups and downs
The pattern of our lives moves about like sound.
We can make frowns when we're sad,
But nothing is ever too bad,
So choose to smile and be happy,
And that's just what you will be.

Never feel alone, just look at a family tree.
You were created by a mom and dad,
Who both had a mom and a dad,
And they too, a mom and dad.
We are a result of a binary.

They make your family tree,
And your own special history,
If here in only spirit,
Call on them, I promise they will hear it.

Be good with your time,
It's a great matter.
Just like matter, time can't be created or destroyed,
Yet around is time's pattern.

Remember remember this special pattern,
Remember everything flows in this interconnected place,
Remember, if nothing else, to keep a smile on your face.

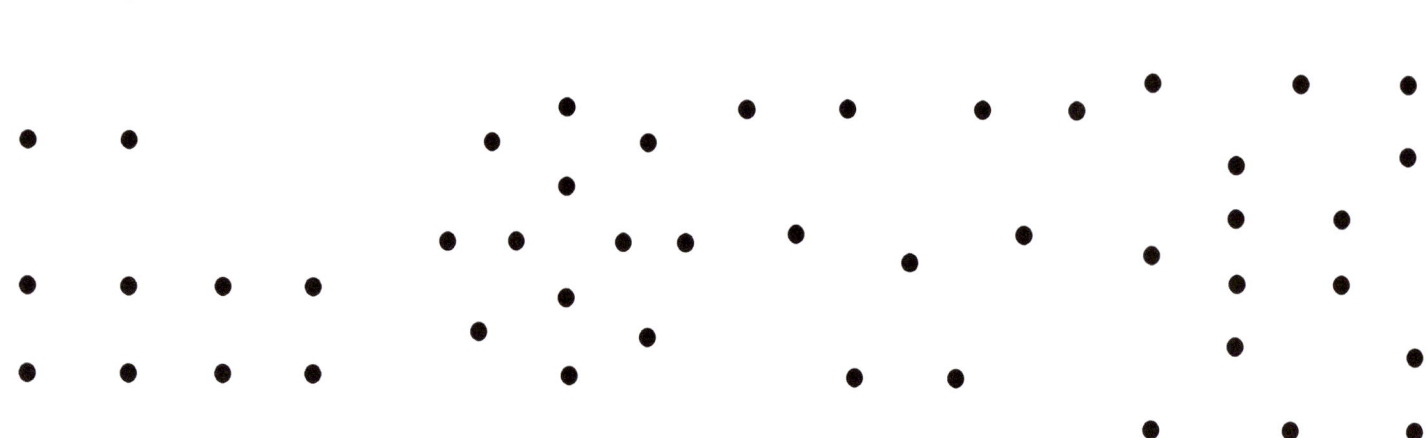

Love as you go, it's an awesome way.
Ever up that staircase of aspiring for higher…
Go share this with your own kids
Who will have their own atoms, molecules, DNA, cells,
Dreams and love to share one day.

WHAT'S THE ATOM TO THE ATOM?

...AS ABOVE, SO BELOW.